CONTENTS

INTRODUCTION TO STREAMING VIDEO

Not too long ago, there weren't many choices to make for television viewers, other than which channel to watch. If you knew which programs you wanted to watch, you could record them and save them for another time; but if you forgot to set up a recording, you could miss it altogether. Sure, you could wait for a rerun, but by then everyone would already be talking about the surprise ending of the season finale.

Or what about that great show that everyone was always talking about around the water cooler? You could try to jump into it in the middle of a season and try to catch up, but wouldn't it be more fun to see it from the beginning?

And most of us don't have unlimited time to watch television—wouldn't it be nice to watch whatever show you want at whatever time you want?

Enter streaming video! Thanks to advancements in data networking and the proliferation of powerful home computers and operating systems, viewers now have many choices for watching their favorite shows, movies, and sporting events.

There are a variety of ways that you can stream video online, whether it be on your smart TV, tablet, computer, or smartphone. Also, there is no lack of production companies providing content for viewers who prefer to stream.

In a nutshell, streaming video is content in a compressed format that is continually transmitted over the internet. Streaming video can be either live or on demand: "live" streaming is often used for content such as sporting events or concerts and is only available for a short time. But "on demand" streaming is saved content that can be played any time the viewer wishes. No more waiting for reruns or needing to turn a show on at a specific time—you can watch any time you want. And what's more, you can jump into a new-to-you series from the very beginning, without missing an episode.

Services like Roku, Netflix, and Hulu are making television and movies more fun than ever. And these services aren't just fun to use, they're simple, too!

With the innovation of high-speed mobile networks and Wi-Fi technology, watching "TV" is no longer something that is restricted to the household.

WHAT IS A SMART TV?

You're probably already using a smartphone—a phone that has internet access and is capable of performing the functions of a computer. But what about a smart TV?

As you may have guessed, smart TVs—like smartphones—have internet access. They also come with built-in Wi-Fi—so wherever you place your smart TV in your home, it should be able to connect to the internet.

Internet access is what makes streaming television shows and movies—often across multiple services like Netflix, Hulu, or YouTube—possible on a smart TV. Many different brands produce smart TVs, all with different operating systems. But they all boast simple-to-use technology, so it's hard to go wrong when choosing a smart TV!

HOW TO USE A SMART TV

Because they can connect to the internet, smart TVs come with features similar to what you might find on your smartphone or tablet. Most smart TVs have a dashboard with apps like Pandora, Netflix, Facebook, or the Weather Channel. So if you have a Netflix or Pandora account, you can easily use these services right on your TV. For instance, you can stream a movie from Netflix without having to rent one. Or you can play music on Pandora right from your television—perfect for setting the mood when you have parties or family over.

The features available with smart TVs vary by brand. Some of them are capable of playing 3D movies, some respond to voice commands, and some come with special dual-sided remotes that make navigating the apps super easy. Some smart TVs can even learn from your viewing habits,

Brand	Operating System	The Lowdown
Samsung	Tizen	Tizen is extremely fast, and it'll automatically detect devices that you connect to the TV, labeling inputs accordingly. Plus you can control connected devices with the TV remote.
LG	WebOS	WebOS is extremely simple and fun to use, and it supports motion controls with the included remote.
Sony	Android TV	If you use an Android phone, this should be immediately familiar. Sony smart TVs support Google Cast, with which you can project your phone's (or tablet's) display onto your TV.
TCL	Roku	Roku OS is awesome, featuring clean, simple navigation and a best-in-class search function that looks through every app for your chosen content. Also: Voice search!
Element	Amazon Fire TV	In addition to the inclusion of the Amazon Video app, you'll get access to Alexa, a personal assistant who can help navigate your TV *and* control your smart home devices.
Westinghouse	Amazon Fire TV	See above

SMART TV

Smart TV operating systems are still developing in both proprietary and open-source frameworks. Many of them are very versatile, allowing the users to download and run apps, watch on-demand media, and even interact with social media.

and make suggestions for new shows and movies based on what you've previously watched.

But nearly all smart TVs are simple and fun to use. By hitting the ***Menu*** button on your remote control, you are taken to a screen with different options, like viewing shows on demand, or seeing available apps. Simply choose what you'd like to view and peruse your selections. If you don't see what you're looking for, smart TVs always come with a search option. Generally, an alphabetic menu will pop up on your screen, which you can navigate using the **Arrow** icons on your remote. Alternatively, if you have a dual-sided remote with a keyboard, you can type out the title of what you're searching for.

With on-demand content now available whenever the viewer wants, a new trend in TV consumption now allows the viewer to watch an entire season or even series in one sitting without enduring commercial breaks or having to wait for the next episode to come out. This new trend is called "binge watching."

SETTING UP YOUR SMART TV

All smart TVs come ready to be connected to an ethernet cable, which is a cable used to connect devices together within close proximity, such as the computers you use in your home. Most smart TVs are also capable of a wireless connection, but depending on where you are putting your TV, you may prefer a wired connection. For instance, if your smart TV is in the basement but your Wi-Fi router is in an upstairs study, the Wi-Fi connection may not always be stable. An ethernet cable will provide a more stable link and prevent your shows and movies from running slowly or freezing up.

Smart TVs don't affect your television's ability to receive local channels, so if you have cable or satellite now, you can still use them with your smart TV. If you don't have cable or satellite, you'll need an HDTV antenna to receive local broadcasts for free. Many smart-TV owners are choosing to "cut the cord" and forgo cable and satellite service to reduce their monthly bills. If you decide to go this route, but still want to have access to local broadcasts, be sure to look for a smart TV labeled *HDTV* or *Ultra HDTV*. If the TV is described as a "home theater display," this means that it cannot use an antenna to pull in local stations.

Before using your smart TV, you'll need to have a home network set up. It's also recommended that you have a Wi-Fi router that supports 802.11ac, which is the most up-to-date standard in Wi-Fi.

If you do want to continue using cable or satellite, a cable or satellite box will still be needed. And of course, to use the smart TV's apps and to stream content, you'll need an internet connection. It's best to have a high-speed connection if possible, as slower speeds can cause hiccups, especially with video services like Netflix. Slower speeds may still be fine for streaming music from services like Pandora or Spotify.

ULTRA HD RESOLUTIONS

Ultra HD (UHD) is a term used for televisions that support an aspect ratio (the proportional relationship between the screen's display width and height) of 16:9, which is the industry standard for most televisions and computer monitors these days. But the television must also have at least one digital input that can carry video at a minimum resolution of 3840x2160 pixels (8.29 megapixels) to be considered UHD. There are two tiers of video resolution for UHD.

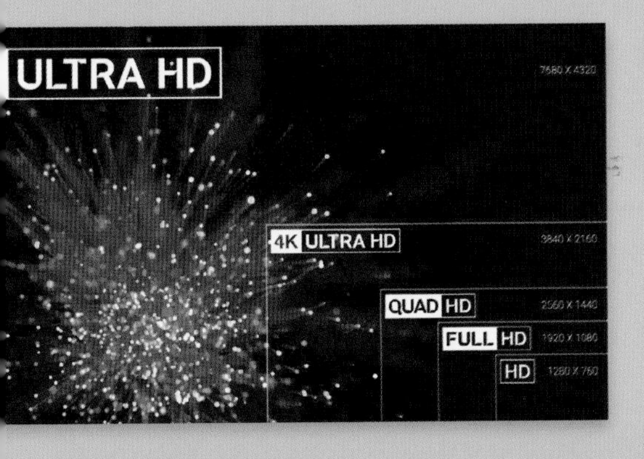

- UHDTV-1 meets the minimum standard of pixel resolution for UHD classification (3840x2160 pixels), but it should not be confused with 4K display. 4K display has a resolution of 4096x2160 pixels. At 8.29 megapixels, UHDTV-1 has four times as many pixels as 1080p HD displays.
- UHDTV-2 has a resolution of 7680x4320 pixels. It is also referred to as 8K UHD and has 16 times as many pixels as 1080p HD display.

WHAT IS ROKU?

A Roku device streams television shows, movies, and music from the internet to your smart TV.

Roku devices—which include Roku Box, Roku Streaming Stick, and Roku TV—come with an operating system that allows users to access whatever content they choose. With up to 4,500 channels, there are plenty of options. Although many of them are free, some channels require additional subscriptions.

Roku devices also come with apps that allow the user to play video and music stored on other PCs or media servers that are connected to your home network.

Setup is easy: just connect your Roku device to your television or power up your Roku TV, then connect to either wired or wireless internet. Roku will provide you with a code to activate your device, then just create an account and start streaming!

If you have a Roku Box or Streaming Stick, you can even take your media with you when you travel. All you need is a television and Wi-Fi—then just plug your Roku into the HDMI port of any TV and log into your account.

Amazon Cloud Player and Video, BBC iPlayer, CBS All Access, DirecTV Now, Google Play Movies & TV, HBO Go, Hulu, MLB.TV, NatGeo TV, NBA TV, Netflix, PBS, Showtime, Time Warner Cable, and Youtube are some of the content providers that can be streamed with Roku.

SEARCHING FOR CONTENT ON ROKU

There are three ways to search for content with Roku: you can search using the on-screen keyboard, you can use your mobile device, or you can simply use your voice. To search with the on-screen keyboard, hit the ***Home*** button, and then navigate to ***Search***. Begin typing the show, movie, actor, director, or channel that you're looking for, and Roku will begin to provide results. Select the one you're looking for to see more details.

To search on your mobile device, you'll need the free Roku mobile app. Open the app, then tap either ***Channels*** or ***What's On*** at the bottom of the screen. Tap the **MAGNIFYING GLASS** icon to access the search page, and then start typing your search. When you see what you're looking for, tap the entry to see more details.

To search with your voice, you must have a Roku Enhanced Remote with voice search. If you don't already have a remote that supports voice search, they can be purchased at Roku.com in the accessories store. To use voice commands with the remote, hold the remote face up, about six inches from your mouth. Press the **MICROPHONE** icon or **MAGNIFYING GLASS** icon, and then speak your command. Once you release the button, Roku will carry out your request. You can ask for titles of movies or shows ("show me episodes of *NCIS*"), specific actors ("find Tom Hanks movies"), or genres ("show me comedies").

ADDING TO *MY FEED*

Because of the large array of content created by a number of providers on Roku, it may be difficult to keep track of your favorite shows between providers. You can make this task more manageable by adding your favorites to the My Feed menu option. Simply search for what you want to keep track of, click on the title, and then select ***Follow on Roku***. The titles you have selected will then be available for easier access in the My Feed menu option.

WHAT IS AMAZON FIRE TV?

Amazon Fire TV is a streaming media player that lets you watch movies and television shows and listen to music from the internet on your TV. Fire TV is integrated with Amazon's own streaming service. You'll need to have accounts with whichever services you want to use with your Amazon Fire TV, or have a cable or satellite package that includes the specific networks you want to watch.

All models of Amazon Fire TV come with Alexa, Amazon's voice-activated "personal assistant." You can pair your Fire TV with your Amazon Echo for hands-free voice commands.

Simply plug your Amazon Fire TV into your HDTV and you'll be streaming video in no time. It's easy.

With Amazon Fire TV you can watch content from a number of providers, including Hulu Plus, Netflix, Watch ESPN, Sling Television, NBC News, PBS Kids, HBO Now, Showtime Anytime, and many others.

SETTING UP YOUR AMAZON FIRE TV

Before you can set up your Amazon Fire TV, you'll need an Amazon account. If you don't already have one, there will be an option to create one during the setup process. You'll also need a high-definition TV with an HDMI input port, and an HDMI cable. And of course, an internet connection!

To get started, connect your Amazon Fire TV device to the HDMI port on your TV, and plug the device into a power source. Make sure your TV is tuned to the right input source—you should see a "Fire TV" screen. Now, click the **PLAY/PAUSE** icon on the remote, which will prompt the system to scan for wireless networks. Choose your home network and enter your password.

Once Fire TV has a connection, you can either create an Amazon account, or enter your current account information. Now Fire TV will play a short video to help you learn the key features of your new system—you're ready to start streaming!

NAVIGATING AMAZON FIRE

To use Amazon Fire TV, you'll use your remote control to navigate through on-screen menus to choose which shows or movies to watch. The **HOME** icon on the remote takes you to the home screen, where you can see recently viewed items, featured movies and shows that Amazon recommends, and new releases. On the left side of the home screen are additional menus for content you've already purchased in your video library.

Use the **SCROLL WHEEL** in the middle of the remote to scroll up, down, left, or right. When you've highlighted the selection you want, hit the button in the center of the **SCROLL WHEEL** to select it.

Users of Amazon Fire TV can also use the console to play video games with the console's provided remote, your smart phone, or with a game controller.

BROWSING AND SAVING VIDEOS

Some Fire TV remotes also come with voice activation, indicated with the **MICROPHONE** icon. If you have this feature on your remote, you can press and hold the **MICROPHONE** icon and say what you're searching for.

To access your content on Fire TV, select ***Your Videos*** from the menu. Your current rentals and purchases will be listed under ***Your Video Library***. Amazon Fire TV comes with a free 30-day Amazon Prime membership, so you can try it out if haven't already. If you're a Prime member, the selections listed in the menus will be customized to show titles available with Prime.

ADDING TO YOUR WATCHLIST

If you see a show or movie you think you'd like to watch eventually, you can add it to your Watchlist. Simply choose ***Add to Watchlist*** in the video details. You won't be able to view these titles until you buy or rent them—if you've already purchased a title, you'll see a ***Watch Now*** option in the video details.

WHAT IS CHROMECAST?

Google's Chromecast is an affordable device that plugs into your TV's HDMI port and allows access to streaming services like Netflix and Hulu, using your smartphone or computer as a remote control. A regular Chromecast is just $35, and Chromecast Ultra is available for $70. The two devices are identical in terms of channel selection and functionality, but the Ultra broadcasts content at a higher resolution and with advanced color technology, giving you a clearer picture.

The device acts as a transmitter between your television and the device you are streaming video on, sort of like an antenna. The device you are using to watch content will send the content to your Chromecast, and then your Chromecast displays it on your television. This allows you to sever your reliance on your smartphone or laptop's small screen and to watch your content on a larger screen.

HOOKING UP CHROMECAST

Your Chromecast will come with cords and power adapters to make setup simple. Just plug one end of the USB power cable into the Chromecast, and the other end into the power supply. Then plug the Chromecast into an HDMI input on your TV, and plug the power supply into an outlet.

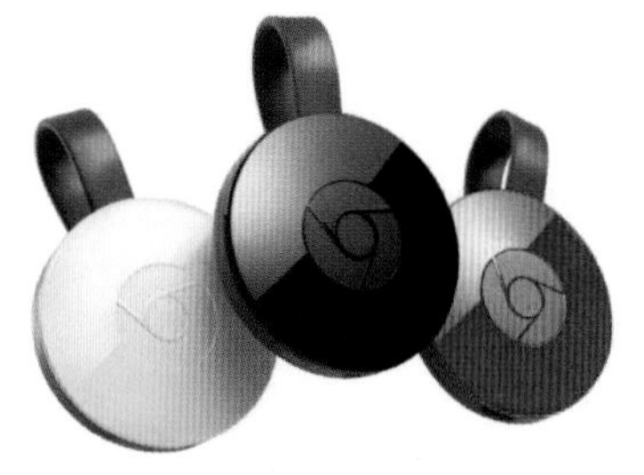

If you'll be using your smartphone as a remote, download the Google Home app, which is available through the Google Play Store or the Apple App Store. If you'll be using your computer, you just need Google Chrome—go to Google's Chromecast website and follow the instructions to get started.

TETHERING DEVICES

The beauty of Chromecast is that it pretty much works as a tethering device between your phone or laptop's app and your television, allowing you to watch a variety of content from Chromecast supported apps. The compatibility ranges from social media apps to video streaming apps to internet radio apps. The content available is truly vast.

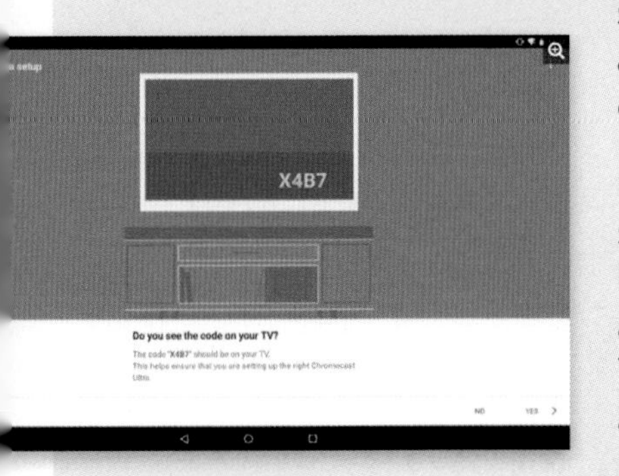

From the Google Home app, select ***Devices*** in the upper right-hand corner. The Google Home app will set up Chromecast for you. You'll see a code on your TV screen and in the app—if they match, select ***Yes***. You can then choose a name for your Chromecast and adjust privacy options. Now, choose the network you want to use and connect your Chromecast to the internet.

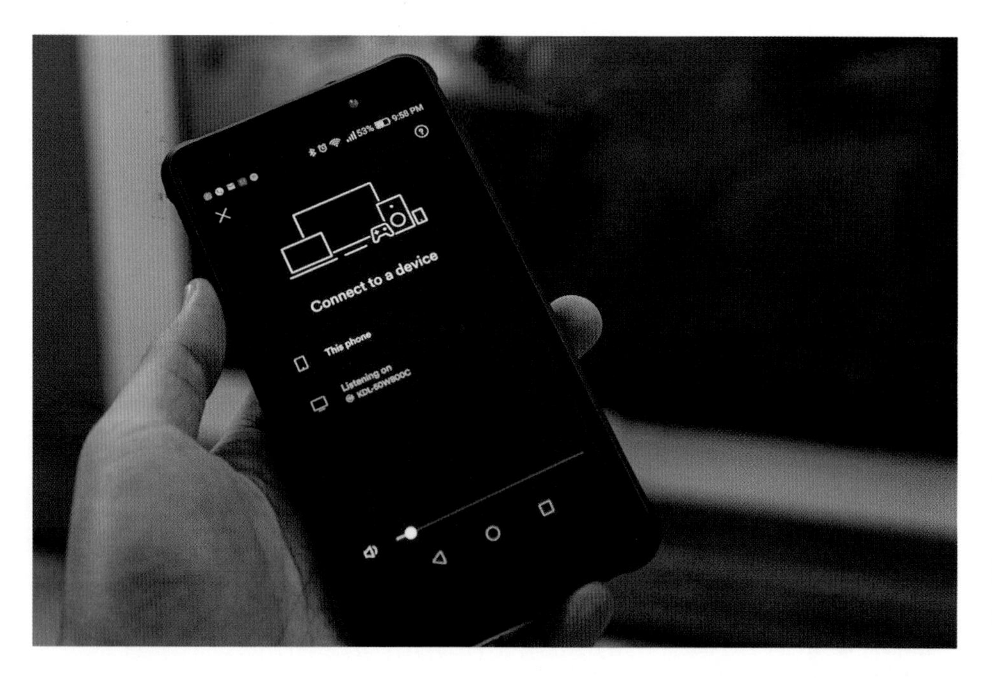

HOW TO USE CHROMECAST

The Google Chromecast website provides a list of apps that can be played through your Chromecast device. These include services like Netflix and Hulu, movie channels like HBO and Showtime, and networks like ABC, NBC, CBS, and Fox.

Sending video to your TV from your Chromecast is called Casting, and it's a simple process. If you're using a computer, just go to your Chrome browser and then go to whichever website you'd like to broadcast from. In the upper right-hand corner of your computer screen, there will be a square-shaped icon—this is the **CAST** icon, and when you click it, it will display the name of your Chromecast device. Click on the device name, and the image on your computer screen will Cast to your television. You can now simply choose whatever video, show, or movie you want to watch and hit play. You can actually broadcast anything at all from your computer to your television using the **CAST** icon, as long as you're using Google Chrome.

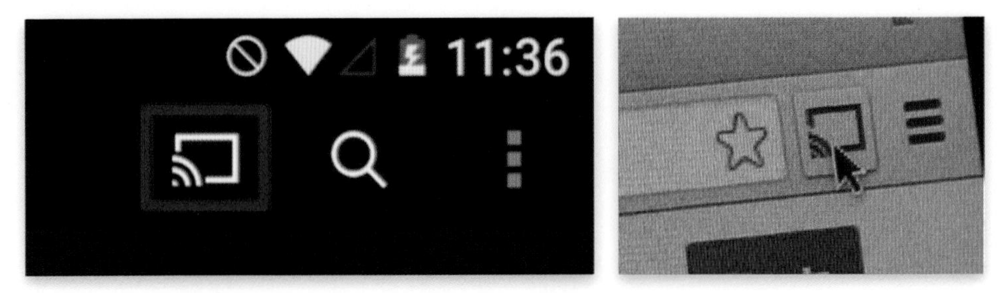

The CAST icon as seen from the mobile Netflix app and the Google Chrome web browser.

Just as when you use your computer to control Chromecast, using your smartphone is simple, as well. Just open a cast-enabled app—for instance, Netflix—and find a video you'd like to play. Then tap the **CAST** icon in the top right-hand corner and control the video from your phone.

Remember, you'll need to have subscriptions to any of the services you use to stream video, as Chromecast merely broadcasts the content for you.

Google Chromecast has been given very positive reviews, especially for its ability to integrate new apps into its Casting network.

WHAT IS NETFLIX?

Of all the video streaming services out there, Netflix may be the most recognizable. And for good reason: the entertainment company has been around for more than 20 years. Originally focusing on DVD rentals and sales, Netflix recognized the immense potential of streaming media and began to expand its services in 2007. Today, the company's streaming videos are available in almost every country in the world!

Netflix is known not only for its streaming videos, but also for its original content. The service produced 126 original series in 2016 alone—more than any network or cable channel. Some of the most popular include *Orange Is the New Black*, *Stranger Things*, and the Marvel superhero series *Daredevil*.

SIGNING UP FOR NETFLIX

The first season of *Stanger Things* is reported as the third most-watched Netflix original series behind the first season of *Fuller House* and the fourth season of *Orange Is the New Black.* It reportedly drew in more than 14 million adults between the ages of 18–49 within the first 35 days.

Netflix offers a free month of services, so to sign up, go to the Netflix home page and click on ***Join Free for a Month***. You'll be given three plan options—which you choose will mostly depend on whether you want HD video, and whether you want to be able to watch on more than one screen. Choose the option you'd like, then continue to step two.

You'll need to enter an email address and a password, and then you'll move on to step three: payment. Even though you'll need to enter your credit card or PayPal information, you won't be charged until your free trial has ended. When you've entered all the relevant information, click the ***Start Membership*** button. You're now ready to enjoy Netflix!

Netflix has international distribution with both its DVD rental-by-mail service and video streaming service in nearly 190 countries. In 2013 it began producing its own content with its debut series *House of Cards.*

SEARCHING FOR CONTENT ON NETFLIX

You can either browse Netflix titles on your TV or on your computer. Some smart TVs will only show you categories that match your viewing habits, so if you want to see every title available, searching on your computer may show you more options. Pull down the ***Browse*** menu to see categories including Action, Classics, Drama, and Musicals. You'll also see extra categories on the left-hand side like Originals—where you'll find Netflix's own shows—and New Arrivals.

To search for something specific, click the **MAGNIFYING GLASS** icon at the top of the page and simply type your search query. You can search for actors, titles, or genres. If you feel like watching a film with a specific actor but aren't sure which one, you can simply type a name—like George Clooney—and Netflix will show you all the titles available starring George Clooney.

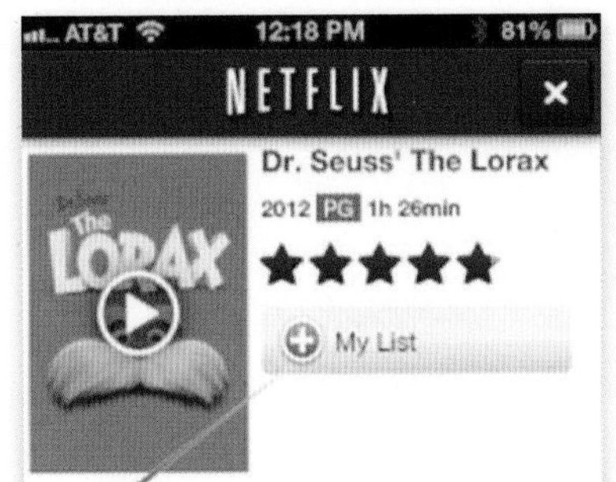

ADDING TO MY LIST

To save movies and shows for later, you can use the My List feature in Netflix. If you're browsing from a smartphone or tablet, tap the Netflix app to open it. Search for the title of the movie or show you want to watch and tap on it. Underneath the title will be a button labeled ***+My List***. Simply tap the button, and the title will be added to your My List selections.

If you're using a computer, hover your mouse over the show or movie you want to select, and tap the **PLUS SIGN (+)** option, or select the title and then choose ***+My List*** to add it.

You can find Netflix Original series and a plethora of classic TV shows and movies in each category.

hulu
LG

WHAT IS HULU?

Hulu is a subscription video-on-demand service that focuses mostly on television series, both past and present. So if you've missed past seasons of certain shows, Hulu may be a great place to catch up on what you missed.

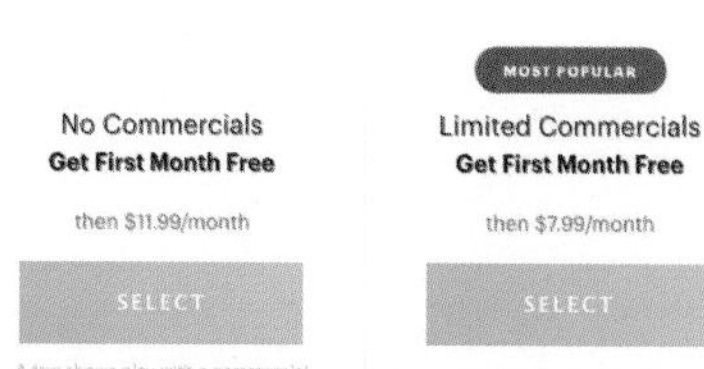
Choose Your TV Experience
Switch plans or cancel anytime.
Hulu
Watch full seasons of exclusive series, current episodes, classic favorites, Hulu Originals, hit movies, kids shows, and tons more with minimal or no commercial interruptions.
MOST POPULAR
No Commercials
Get First Month Free
then $11.99/month
SELECT
Limited Commercials
Get First Month Free
then $7.99/month
SELECT
A few shows play with a commercial before and after the video. Learn More

Hulu might be unique among streaming services in that it generates revenue via subscription fees and advertising content.

HULU

Hulu is more oriented toward TV shows than some other streaming services. It provides content from channels like A&E, FYI, History, Lifetime, Viceland, AMC, Sundance, BBC America, CBS, Animal Planet, Discovery, Fox, Cinemax, Showtime, Adult Swim, Cartoon Network, TNT, TBS, and many others.

SIGNING UP FOR HULU

To sign up for Hulu, go to Hulu.com and click on *Start Your Free Trial*. Then choose the plan you'd like—you can either watch shows on Hulu with limited commercials, or, for an extra fee, choose to watch commercial free.

You can then create an account either using your Facebook information, or with an email and password. Fill in the information, and hit *Continue*. You'll then be asked to enter your credit card or PayPal information—you won't be charged until your free trial ends. Then just hit *Start My Free Trial*, and you're ready to use Hulu!

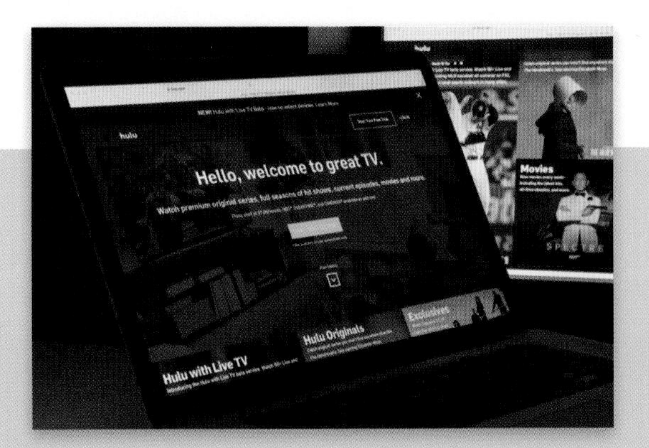

Hulu has also thrown its name into the original programing ring, with shows like *The Path* and *The Handmaid's Tale*.

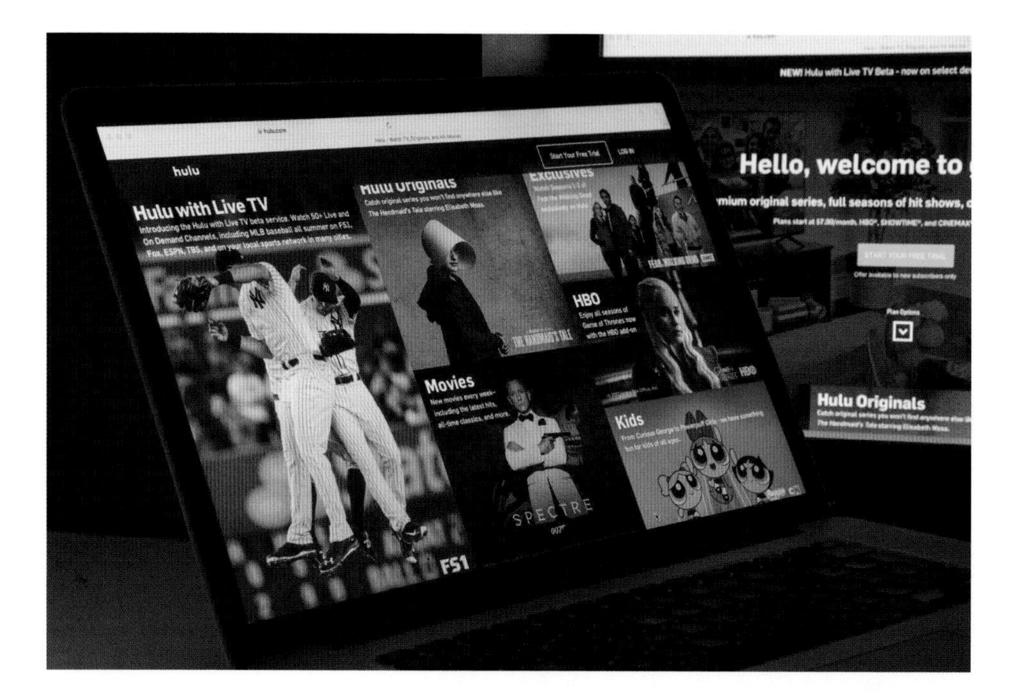

WATCHING HULU ON SMART TVs

You can watch Hulu shows and videos on your TV or your computer. To watch on your television, you'll need a device like Chromecast, Amazon Fire TV, or Roku. If you're using Chromecast, you simply hit the CAST icon to broadcast your show to your TV. If you're using an Amazon Fire TV or Roku, you'll just need to add the Hulu channel to your device to watch content.

It will be easy for you to watch Hulu on your computer, but you may need a streaming device or Hulu-enable smart TV to watch Hulu content on your television. Roku is an easy and cheap way to solve this problem.

Some smart TVs have applications, like Hulu, built right in. If your TV came with a Hulu app, all you have to do is log into your account and start streaming!

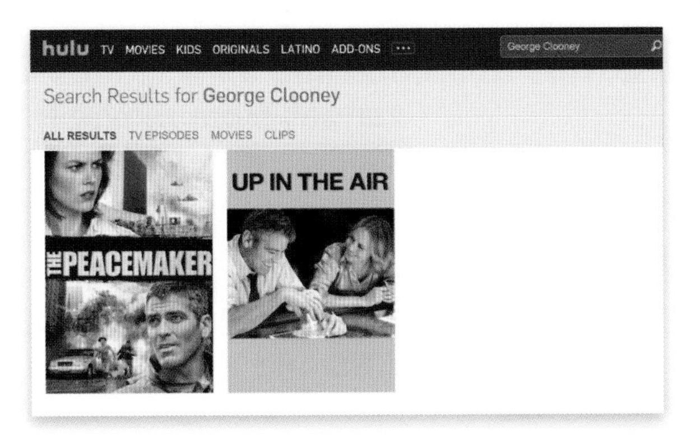

You can search Hulu's content by TV show or movie title, actor name, or episode title.

HULU ON YOUR COMPUTER

To watch Hulu on your computer, simply go to Hulu.com and log in. You can browse titles, or type what you're looking for into the *Search* field at the top of the screen. When you find a movie or show you want to watch, hover over the title, and hit the PLAY icon. Then just grab the popcorn and start watching as your selection begins to play!

DEVICES FOR WATCHING HULU

You don't have to watch Hulu on your smart TV or even at home if you don't want to. There are a variety of devices that you can download the Hulu app onto to watch your content, including Android phones and tablets, Fire tablets, iPhones and iPads, and Nintendo Switch.

WHAT IS YOUTUBE?

Founded in 2005, YouTube is a video sharing website that allows users to upload and view videos. Content ranges from movie trailers and music videos to educational videos and documentary films. Most of YouTube's content is uploaded by individuals, but some media outlets, such as CBS and Hulu, partner with the site to offer content to YouTube viewers.

Anyone can watch videos on the site but you need to be registered in order to upload videos or comment on videos. With more than 400 hours of content uploaded to YouTube every minute, there's always something different to watch!

If you only want to watch videos on YouTube, you don't need to sign in. Simply search for whatever you're looking for—"*Star Wars* movie trailers" for instance—and YouTube will show you a list of results. You can filter your results by upload dates, type, duration, and features by using the *Filter* button toward the right of your browser window.

SIGNING UP FOR YOUTUBE

If you do want to sign in, all you have to do is use a Google account—if you have a Gmail address, you already have a Google account. If not, just go to Google.com and click the ***Sign Up*** button. Fill out the required information, and you're good to go!

When you sign in to YouTube, you have the option of sharing videos through email or social media, signing up for subscriptions to be notified when new videos are posted, or saving videos to view at a later time.

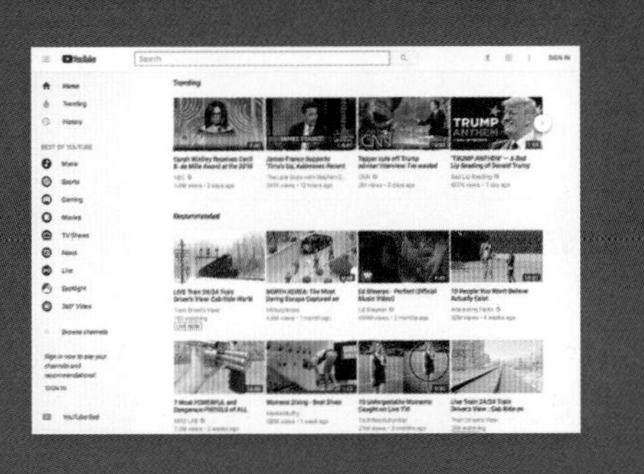

YouTube offers a humungous variety of content from an array of independent and syndicated channels. You can find music videos, news, TV shows (for a payment), live-streaming entertainment, vlogs, and an endless mix of YouTube personalities talking about anything and everything. Here we can see videos uploaded by CNN, NBC, and a host of independent channels. There is even a Live Feed of a train driver's view being broadcast with 763 people watching.

If you find a channel that you like and want to be kept up to date about what they are posting, hit the red *Subscribe* button at the bottom-right of the video frame to receive notifications whenever they upload new videos.

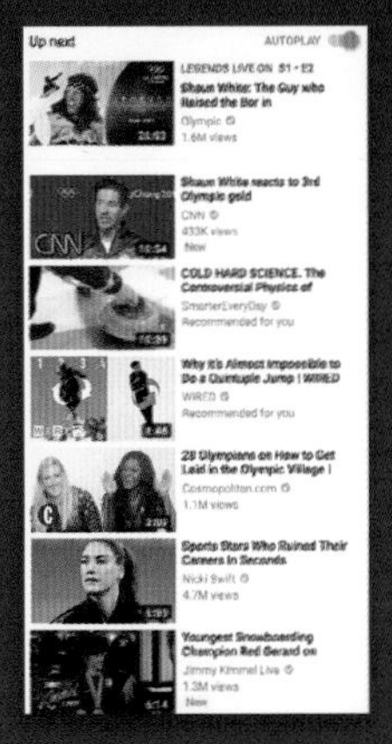

ADDING VIDEOS TO PLAYLISTS

If you want to watch a video that is queued in the ***Up Next*** sidebar at a later time, hover your mouse over the video and click on the **CLOCK** icon in the upper right-hand corner. When you click this icon, the video will be added to the Watch Later playlist in your library. This is convenient if you know you'll want to watch the video eventually, but don't want to have to search for it again later. You can also add the video you are currently watching to your Watch Later playlist, or any of your other playlists, by tapping the icon directly to the right of the ***Share*** button.

CREATING YOUR OWN CHANNEL ON YOUTUBE

To create your own channel on YouTube, sign in, click on your **PROFILE** icon in the top right-hand corner, and then click on ***Settings***. Next to your email address in the account information, click on ***Create a Channel***. You can use your own name, a business name, or create a new name for your channel. That's it! You can now customize the photos on your channel by clicking on the **PENCIL** icon in the corner of your profile picture, and clicking ***Add Channel Art*** in the center of the screen.

From left to right you have your UPLOAD, APPLICATIONS, and NOTIFICATIONS icons. On the far right is your PROFILE icon.